NATIONAL GEOGRAPHIC KIDS

美 国 国 家 地 理
双 语 阅 读

Volcanoes
火山

懿海文化 编著

马鸣 译

第三级

外语教学与研究出版社
FOREIGN LANGUAGE TEACHING AND RESEARCH PRESS
北京 BEIJING

京权图字：01-2021-5130

图书在版编目（CIP）数据

火山：英文、汉文／懿海文化编著；马鸣译. −− 北京：外语教学与研究出版社，2021.11（2023.8 重印）
（美国国家地理双语阅读. 第三级）
书名原文：Volcanoes
ISBN 978-7-5213-3147-9

Ⅰ. ①火… Ⅱ. ①懿… ②马… Ⅲ. ①火山－少儿读物－英、汉 Ⅳ. ①P317−49

中国版本图书馆 CIP 数据核字 (2021) 第 236729 号

出 版 人　王　芳
策划编辑　许海峰　刘秀玲　姚　璐
责任编辑　姚　璐
责任校对　华　蕾
装帧设计　许　岚
出版发行　外语教学与研究出版社
社　　址　北京市西三环北路 19 号（100089）
网　　址　https://www.fltrp.com
印　　刷　天津海顺印业包装有限公司
开　　本　650×980　1/16
印　　张　37.5
版　　次　2022 年 3 月第 1 版　2023 年 8 月第 4 次印刷
书　　号　ISBN 978-7-5213-3147-9
定　　价　188.00 元（全 15 册）

如有图书采购需求，图书内容或印刷装订等问题，侵权、盗版书籍等线索，请拨打以下电话或关注官方服务号：
客服电话：400 898 7008
官方服务号：微信搜索并关注公众号"外研社官方服务号"
外研社购书网址：https://fltrp.tmall.com

物料号：331470001

Table of Contents

Mountains of Fire!

Ash and steam pour out of the mountain. Hot melted rock rises up inside the mountain. Suddenly a spray of glowing hot ash shoots out. It is an eruption!

More melted rock is forced out. It spills down the side of the volcano in a burning hot river. Anything that cannot move is burned or buried.

KIMANURA VOLCANO
ZAIRE

WORD BLAST

ERUPTION: When magma reaches Earth's surface. Some eruptions are explosive.

Hot Rocks

When magma comes out of the Earth it is called lava.
The lava hardens.
Ash and rock pile up.
A volcano is born.

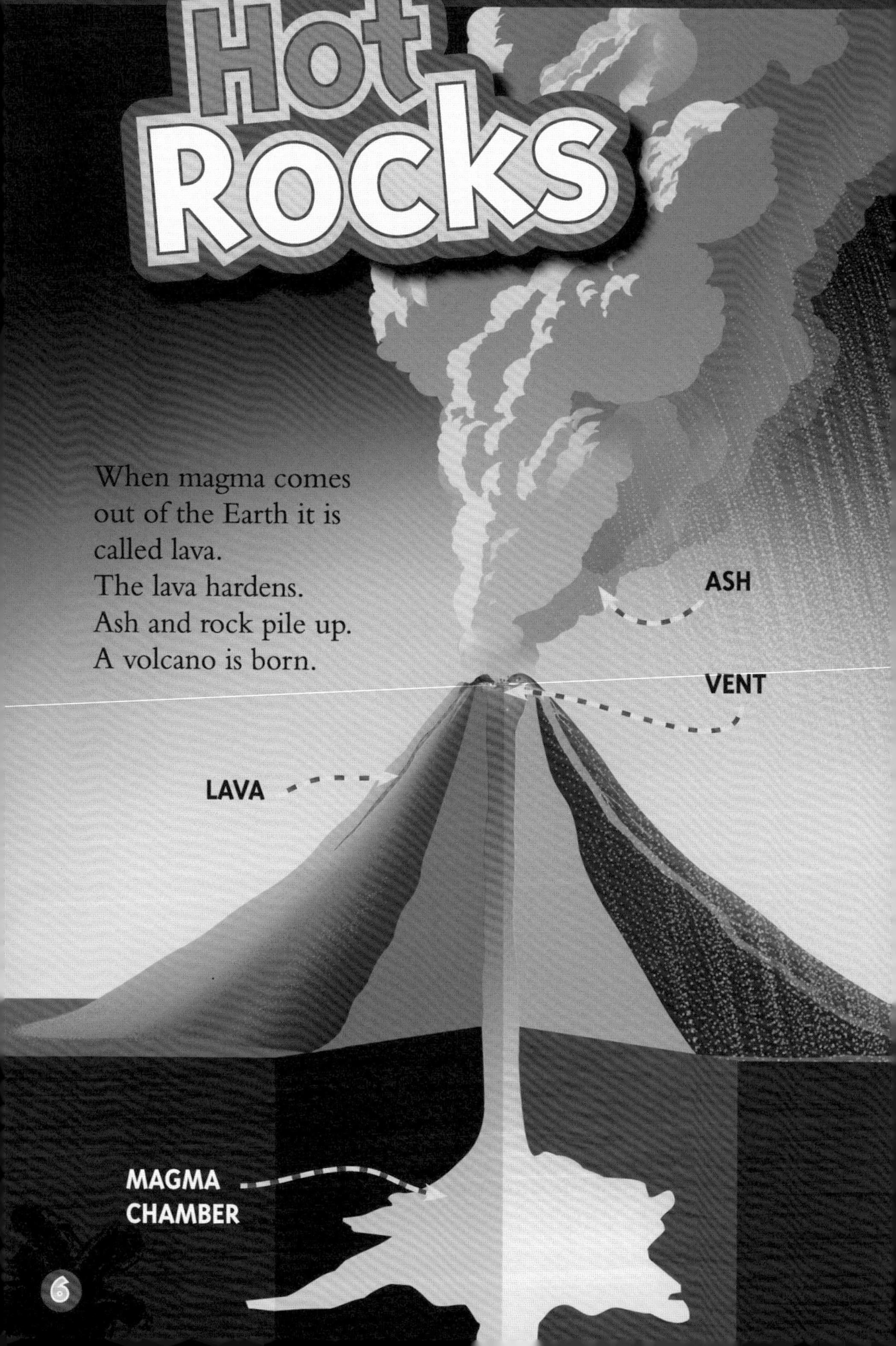

ASH

VENT

LAVA

MAGMA CHAMBER

Deep beneath the Earth's surface it is hot. Hot enough to melt rock. When rock melts it becomes a thick liquid called magma. Sometimes it puddles together in a magma chamber. Sometimes it finds cracks to travel through. If magma travels through a crack to the surface, the place it comes out is called a vent.

WORDS BLAST

MAGMA: Thick, liquid melted rock

MAGMA CHAMBER: A space deep underground filled with melted rock

VENT: Any opening in Earth's surface where volcanic materials come out

Shaky Plates

Where do cracks and vents
in the Earth come from?

The land we live on is broken into pieces
called plates. The plates fit Earth like a
puzzle. They are always moving a few
inches a year. When plates pull apart...
or smash together...watch out!

This picture shows the gap that forms when plates pull apart.

THINGVELLIR, ICELAND

One place where Earth's plates smash together is called the Mid-Atlantic Ridge. It is the longest mountain range on Earth and most of it is underwater.

An Island Is Born

What happens when two plates pull apart?
They make a giant crack in the Earth.
Magma can rise up through these cracks.
This even happens underwater.

About 60 million years ago an underwater volcano poured out so much lava, it made new land. A huge island grew, right in the middle of the ocean. Lava formed the country of Iceland!

SURTSEY

About 50 years ago, people saw smoke coming out of the ocean near Iceland. A new island was being born right before their eyes! They called it Surtsey, after the Norse god of fire.

The Ring of Fire

What happens when plates bump into each other? Maybe a mountain will be pushed a little higher. Maybe a volcano will erupt. There may be an earthquake, or a tsunami, or both!

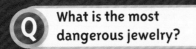
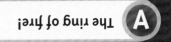
WORD BLAST

TSUNAMI: Large waves created by events like earthquakes and landslides

The edge of the Pacific plate is grinding into the plates around it. The area is called the Ring of Fire. Many of Earth's earthquakes and volcanoes happen in the Ring of Fire.

Postcards from the Ring

I Lava YOU!

Mount Merapi, Indonesia

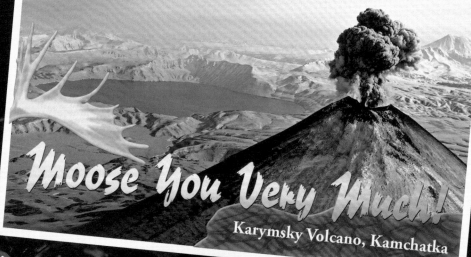

Moose You Very Much!

Karymsky Volcano, Kamchatka

Washing You A Great Day!

From the Cascade Mountains in Washington State

Mount St. Helens

HAVING A HOT TIME IN THE ANDES!

Tungurahua Volcano, Ecuador

Meet a Volcano... Or Three

Not all volcanoes are the same. What kind they are depends on how they erupt.

The lava from a shield volcano is hot and liquid. Rivers of lava flow from the volcano's vents. These lava flows create a gently sloping volcano.

HOT FACT

Olympus Mons on Mars is a shield volcano. It is the largest volcano in our solar system! Seen from above, it is round, like a shield.

MAUNA LOA

The Hawaiian myth of Pele tells the story of how Pele, goddess of earth and fire, built a home on Mauna Loa. Violent volcanic eruptions are said to be Pele losing her temper.

Meet Mauna Loa!

PARICUTIN VOLCANO

A cone volcano has straight sides and tall, steep slopes. These volcanoes have beautiful eruptions. Hot ash and rock shoot high into the air. Lava flows from the cone.

One day a cone volcano started erupting in a field in Mexico. It erupted for nine years. When it stopped it was almost as high as the Empire State Building.

Meet Paricutin!

HOT FACT Even though Paricutin stopped exploding in 1952, the ground around it is still hot! Scientists guess that Paricutin spit out 10 trillion pounds of ash and rock.

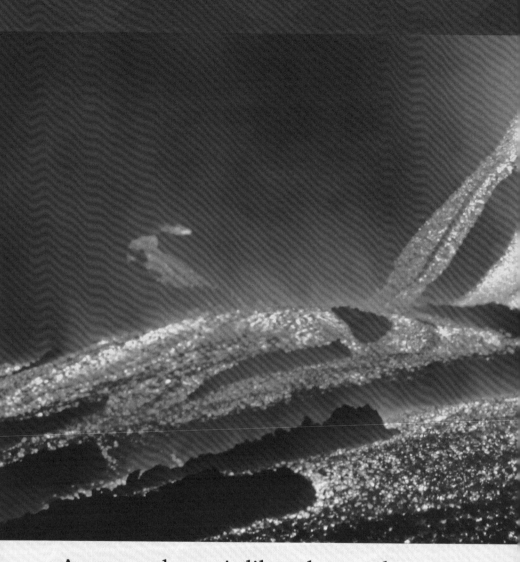

A stratovolcano is like a layer cake.
First, lava shoots out and coats the
mountain. Then come rock and ash.
Then, more lava. The mountain builds
up with layers of lava, rock, and ash.

Meet Mount Etna!

There is a myth about Vulcan, a Roman god of fire and iron. He lived under Vulcano Island, near Mount Etna. Every time Vulcan pounded his hammer, a volcano erupted. The word *volcano* comes from the name Vulcan.

The True Story of Crater Lake

Crater Lake may seem like a regular lake, but it is actually a stratovolcano. It was once a mountain called Mount Mazama. Now it is a deep, clear lake in Oregon.

An explosion over 6,000 years ago blew the top off Mount Mazama. Lava, dust, and ash swept down the mountain. The mountaintop fell in and a giant caldera was formed. Over time the caldera, a crater, filled with water. It is the deepest lake in the United States.

WORD BLAST

CALDERA: A caldera is formed when the top of a volcano caves in.

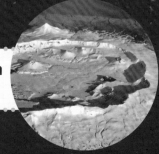

CRATER LAKE

After the mountain collapsed, there were more eruptions. In one, a small cinder cone of ash and lava was formed. This cinder cone pokes out of the lake. It is called Wizard Island.

Volcanoes Rock!

PAHOEHOE

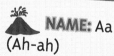 **NAME:** Pahoehoe (Pa-hoy-hoy)

HOW IT FORMS: Fast, hot, liquid lava hardens into smooth rope-like rock.

SPECIAL POWER: It hardens into beautiful and weird shapes known as Lava Sculptures.

AA

NAME: Aa (Ah-ah)

HOW IT FORMS: The crust on top of Aa lava hardens into sharp mounds of rocks.

SPECIAL POWER: It can cut right through the bottom of your shoes!

PELE'S HAIR

🌋 **NAME:** Pele's Hair (Pel-lay)

🌋 **HOW IT FORMS:** Lava fountains throw lava into the air where small bits stretch out and form glass threads.

🌋 **SPECIAL POWER:** These strands of volcanic glass are super thin and long, just like hair! Small tear-shaped pieces of glass, called Pele's tears, sometimes form at the end of Pele's Hair.

PUMICE

🌋 **NAME:** Pumice (Puh-miss)

🌋 **HOW IT FORMS:** In a big explosion, molten rock can get filled with gas from the volcano. When the lava hardens the gas is trapped inside.

🌋 **SPECIAL POWER:** The gas makes the rock so light, it can float on water.

Volcanic Record Breakers

Indonesia, a string of islands in the Ring of Fire, has **more erupting volcanoes** than anywhere else on Earth.

JAVA ISLAND

Q What did the astronomer say about the volcanoes on Io?

A "They're out of this world!"

The place with the **most volcanic activity** is not on Earth. It is on Io, one of Jupiter's moons!

The 1883 explosion of Krakatau was the **loudest sound** in recorded time. People heard the explosion over 2,500 miles away. Anak Krakatau, which means "Child of Krakatau," is a volcano that was born in 1927 where Krakatau used to be.

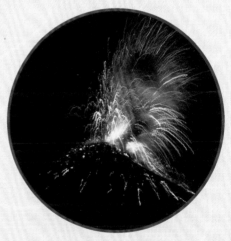

Mount Etna is the **largest active volcano** in Europe.

Hot Spots

Do you want to visit somewhere really hot? Check out these hot spots—places on Earth where magma finds its way through the Earth's crust. Hot spots are heated by volcanic activity!

The Hawaiian Islands are all volcanic mountains. They start on the sea floor and poke out above the sea. Kilauea in Hawai'i is still erupting. As long as it keeps erupting the island of Hawai'i keeps growing.

On Kyushu Island, in Japan, some people use the hot springs to boil their eggs.

Take a bath with the monkeys in Japan.

In Iceland, you can swim in pools heated by volcanoes.

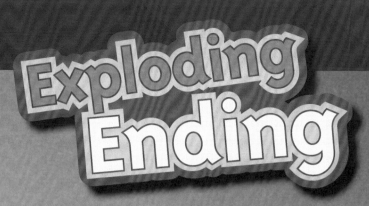

Exploding Ending

If you visit Yellowstone National Park, you will be standing on one of the biggest supervolcanoes on Earth. Yellowstone sits on an ancient caldera. Magma still bubbles and boils a few miles below ground.

Yellowstone has a lot of geysers—more than 500. The magma below Yellowstone caldera heats underground water. The water boils and bursts to the surface as geysers, spraying steam and hot water high into the air.

Go to Yellowstone and see Earth in action!

Glossary

CALDERA: A caldera is formed when the top of a volcano caves in.

ERUPTION: When magma reaches Earth's surface.Some eruptions are explosive.

MAGMA: Thick, liquid melted rock

MAGMA CHAMBER: A space deep underground filled with melted rock

TSUNAMI: Large waves created by events like earthquakes and landslides

VENT: Any opening in Earth's surface where volcanic materials come out

▶ 第 4—5 页

火之山！

火山灰和热浪从山体里喷出来。在山体里，炽热的熔岩不断上涌。突然，炽热的火山灰喷涌而出。火山喷发了！

更多的熔岩喷出来。它沿着火山侧面向下流，汇成一条燃烧着的炽热的河流。任何不能移动的东西都被烧毁或掩埋。

火山小词典

火山喷发：当岩浆到达地表的时候。有些火山喷发是爆炸性的。

基穆拉火山
扎伊尔

▶ 第 6—7 页

炽热的岩石

岩浆从地球内部喷涌而出后被叫作"火山岩浆"。

火山岩浆逐渐变硬。火山灰和岩石不断堆积，一座火山就形成了。

火山岩浆

岩浆房

火山灰

喷出口

地表之下的深处非常热，温度高到可以将岩石熔化。岩石熔化后形成的浓稠液体叫"岩浆"。有时它会汇聚在一起，形成岩浆房。有时它会顺着裂隙流动。如果岩浆顺着裂隙流到了地表，喷出的位置被叫作"喷出口"。

火山小词典

岩浆：浓稠的液态熔岩

岩浆房：地球深处熔岩聚集的地方

喷出口：火山质喷发出来的地表缺口

▶ 第8—9页

摇晃的板块

地球上的裂隙和喷出口是怎么形成的呢?

我们居住的陆地分裂为好多块,被称为"板块"。这些板块像拼图一样拼成了地球。它们每年都会移动几英寸(1英寸约等于2.54厘米)。如果板块拉伸分裂……或者撞到一起……当心啊!

这张图展示的是板块拉伸分裂时形成的裂隙。

辛格韦德利,冰岛

地球的板块相撞形成的这个地方叫"大西洋中脊"。它是地球上最长的山脉,它的大部分都在水下。

▶ 第10—11页

岛屿诞生了

如果两个板块拉伸分裂,会发生什么呢? 它们在地球上形成一个巨大的裂隙。岩浆会从裂隙中喷出来。这种情况即使在水下也会发生。

大约6,000万年前,一座水下火山喷出很多火山岩浆,形成了一块新大陆。一座巨大的岛屿在海洋深处拔地而起。火山岩浆形成了冰岛这个国家!

大约50年前,人们看到冰岛附近的海面上升起了浓烟。一座全新的小岛在人们的眼前诞生了! 他们叫它"叙尔特塞",来源于斯堪的纳维亚语里的"火神"。

叙尔特塞

▶ 第 12—13 页

环太平洋火山带

　　如果板块相撞会发生什么呢？可能一座山会在挤压之下变高一些。可能一座火山会喷发。还可能会出现地震或海啸，或者这两者同时发生！

　　太平洋板块的边缘一直与周边的板块发生摩擦，这片区域被叫作"环太平洋火山带"。地球上的很多地震和火山喷发都发生在环太平洋火山带。

火山小词典

海啸：地震或者山崩等现象引发的巨浪

▶ 第 14—15 页

来自环太平洋火山带的明信片

默拉皮火山，印度尼西亚

卡雷姆火山，勘察加半岛

来自华盛顿州的喀斯喀特山

圣海伦斯火山

通古拉瓦火山，厄瓜多尔

认识一座火山……或者三座

　　并非所有的火山都是一样的。它们的喷发方式决定了它们的类别。

　　盾形火山喷发出的火山岩浆是炽热的、液态的。火山岩浆汇成的河流从火山的喷出口流出来。火山岩浆变硬之后形成坡度平缓的火山。

　　夏威夷神话中佩蕾女神的故事讲述了掌管土地和火的佩蕾女神如何在冒纳罗亚火山上建造家园。据说，剧烈的火山喷发是培雷女神在发脾气。

认识一下冒纳罗亚火山吧！

冒纳罗亚火山

疯狂的真相

　　火星上的奥林匹斯山是一座盾形火山。它是太阳系最大的火山！从太空俯视，它就像一个圆形的盾牌。

▶ 第 18—19 页

帕里库廷火山

锥形火山又直又高，非常陡峭。这类火山在喷发时非常好看。炽热的火山灰和岩石喷到高空。火山岩浆从锥体上流下来。

有一天，位于墨西哥的原野上的一座锥形火山喷发了。它持续喷发了9年。当它停止喷发时，它几乎和帝国大厦一样高。

认识一下帕里库廷火山吧！

疯狂的真相

尽管帕里库廷火山于1952年停止了喷发，但是它周围的土壤仍然非常热！科学家推测，帕里库廷火山喷出了大约10万亿磅（约4.54万亿千克）的火山灰与岩石。

▶ 第 20—21 页

成层火山就像千层蛋糕。首先，火山岩浆喷出来后包裹住了山体。接着是岩石和火山灰。接着又是火山岩浆。成层火山就是由一层又一层的火山岩浆、岩石和火山灰堆积而成的。

认识一下埃特纳火山吧！

有一个神话是关于罗马神话中的火与锻冶之神伏尔甘的。他住在埃特纳火山附近的武尔卡诺岛下面。每当伏尔甘敲响他的锤头时，一座火山就会喷发。英文单词 volcano 来源于伏尔甘的名字 Vulcan。

埃特纳火山，意大利

▶ 第 22—23 页

火山口湖纪事

火山口湖看上去就像普通的湖，但实际上它是成层火山的一部分。这里曾经是一座叫"梅扎马火山"的高山，现在它是俄勒冈州一座幽深、清澈的湖。

6,000多年前，梅扎马火山发生了一次喷发，火山岩浆、尘埃和火山灰从山上席卷而下，山顶骤然塌陷，形成了一个巨大的破火山口。日积月累，破火山口积满了水，形成了火山口湖。它是美国最深的湖。

火山口湖

火山小词典

破火山口：当火山顶塌陷时，破火山口就形成了。

山体崩塌之后，火山喷发还会继续。火山灰和火山岩浆堆积成一座小型的火山锥。火山锥从湖中凸出来，它被叫作"巫师岛"。

▶ 第 24—25 页

火山岩！

绳状熔岩

名字：绳状熔岩

形成方式：快速流淌的炽热的液态火山岩浆变硬后，形成光滑的绳状岩石。

特点：它变硬后形成美丽而奇异的形状，被称为"火山岩浆雕塑"。

块状熔岩

名字：块状熔岩

形成方式：块状熔岩的火山岩浆外皮变硬后，形成锋利的岩石堆。

特点：它可以刺穿你的鞋底！

火山毛

名字：火山毛

形成方式：火山岩浆喷涌而出，射向空中，细小的火山岩浆四散开，形成玻璃丝。

特点：这些火山玻璃丝非常细、非常长，就像毛发一样！在火山毛的末端有时会有泪痕状的小块玻璃，被称为"佩蕾的长发"。

浮岩

名字：浮岩

形成方式：剧烈的火山喷发时，熔岩会包裹着气体从火山中喷出来。当火山岩浆变硬后，气体便被困在里面。

特点：气体使得这种岩石很轻，可以漂浮在水面上。

▶ 第 26—27 页

火山纪录打破者

爪哇岛

印度尼西亚是环太平洋火山带上的群岛，拥有的活火山比地球上别的地方都多。

火山活动最为活跃的地方不在地球上，而在木卫一上，那是木星的一颗卫星！

1883 年的喀拉喀托火山喷发是有记载以来声音最响的一次。远在 2,500 多英里(约 4,023.36 千米) 以外的人们也听到了爆炸声。阿纳喀拉喀托意为 "喀拉喀托之子"，它是 1927 年在喀拉喀托火山原来的位置上形成的火山。

埃特纳火山是欧洲最大的活火山。

▶ 第 28—29 页

炽热之地

你想去炽热的地方参观吗？查一查这些炽热之地吧——就在地球上，岩浆从这些地方的裂隙中流出来。火山活动为这些炽热之地加热！

夏威夷群岛上到处都是火山。它们从海底开始，一直延伸到海平面以上。夏威夷群岛的基拉韦厄火山仍然在喷发。只要它继续喷发，夏威夷群岛就会继续 "长高"。

在日本和猴子一起洗温泉。

在日本的九州岛上，一些人用温泉煮鸡蛋。

在冰岛，你可以在火山地热池里游泳。

▶ 第 30—31 页

爆炸结束后

如果你到了黄石国家公园，你就登上了地球上最大的超级火山之一。黄石国家公园坐落在古老的破火山口之上。岩浆仍然在地表几英里（1 英里约等于 1.61 千米）之下翻腾着。

黄石国家公园有很多间歇泉——超过 500 个。黄石国家公园的破火山口下面的岩浆给地下水加热。水沸腾后喷出地表形成间歇泉，水雾弥漫，温热的水花四溅。

快去黄石国家公园看一看地球的脉动吧！

词汇表

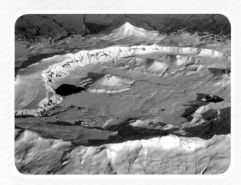

破火山口：当火山顶塌陷时，破火山口就形成了。

火山喷发：当岩浆到达地表的时候。有些火山喷发是爆炸性的。

岩浆：浓稠的液态熔岩

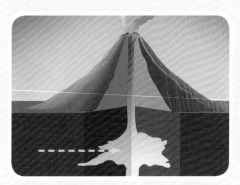

岩浆房：地球深处熔岩聚集的地方

海啸：地震或者山崩等现象引发的巨浪

喷出口：火山质喷发出来的地表缺口